Turner's Bantam Book
Game and Ornamental Bantam Chickens

by G. Archie Turner

with an introduction by Jackson Chambers

This work contains material that was originally published in 1900.

This publication is within the Public Domain.

This edition is reprinted for educational purposes
and in accordance with all applicable Federal Laws.

Self Reliance Books

Get more historic titles on animal and stock breeding, gardening and old fashioned skills by visiting us at:

http://selfreliancebooks.blogspot.com/

Introduction

I am pleased to present yet another title on Poultry.

The work is in the Public Domain and is re-printed here in accordance with Federal Laws.

As with all reprinted books of this age that are intended to perfectly reproduce the original edition, considerable pains and effort had to be undertaken to correct fading and sometimes outright damage to existing proofs of this title. At times, this task is quite monumental, requiring an almost total "rebuilding" of some pages from digital proofs of multiple copies. Despite this, imperfections still sometimes exist in the final proof and may detract from the visual appearance of the text.

I hope you enjoy reading this book as much as I enjoyed making it available to readers again.

Jackson Chambers

Introduction....

※ ※ ※

     BANTAMS have gained great popularity among the Fanciers until to-day hundreds are breeding them exclusively. Year by year new varieties are being added to the already large list of Bantams, and many believe that in the near future nearly all varieties of large breeds of fowls will have their likeness as miniature Bantams. In the past few years so much has been said and written about Bantams that it is almost impossible to say much of them that has not been already stated. Yet the Author's desire is to make this little Book beneficial to its readers. Also, it is his wish that from the articles written by the noted and experienced Bantam Breeders, much interest and benefit may be received.

     Hoping that many more will join with us in the Bantam fancy, I am

       Respectfully yours,

         G. ARCHIE TURNER.

Breeding of Bantams.

ONE of the most important steps to successful Bantam breeding is the housing of the birds. Bantams should have their houses perfectly free from dampness and reasonably warm in winter. Great care should be taken in regard to damp quarters, as dampness causes disease which generally proves fatal. Much care should be used in erecting new buildings, that the lumber should be thoroughly seasoned, as "green" lumber, even put up in early fall, will cause the building to be damp the entire winter. Buildings should always have board floors, thus avoiding dampness. For Ornamental Bantams I should advise to keep closely in-doors during the winter months, while the Game varieties, quite to the contrary, will prosper much better to have free access to a low scratching shed, properly protected from the severe winds and storms. For Bantam Chicks roomy coops should be provided with good ventilation (avoiding draught) and board floors, well sanded, covered with chaff, hay or any dry litter; also made so as to be closed at night to keep out marauding animals. The above I have learned by dearly bought experience. When chicks are about three months old they should be taken from their small coops and put in larger quarters. For this a good dry good box answers the purpose nicely, when a more expensive coop cannot be procured, perches being arranged in box and the floor kept well sanded. By these means coops can be kept clean, droppings being easily removed and birds kept clean and in a healthy and growing condition. Young chicks when very small should never be allowed to run in the wet grass or on damp ground and never expose them to storms, as a Bantam Chick once thoroughly wet will become chilled and very seldom recovers. During the hot summer months the coops should be placed where there is shade if possible, if not artificial shelter may easily be made by driving four stakes and covering them with canvas, oil cloth or boards. This extra work will be highly appreciated by the little creatures. Bantams can be kept nicely in these quarters until the latter part of October or the first of November, according to the weather, when they should be removed to their winter quarters.

PROPER FOOD AND HOW TO FEED.

PROPER food and the feeding of Bantams is of great importance, as they should never be over-fed nor starved. Always give the best of feed obtainable, as they eat a small quantity which should be made up in quality. Feeding twice a day is sufficient in breeding season. During the winter months a light feed at noon, scattered among the litter, keeps the little creatures busy and singing the entire day. During the breeding season, when the birds are confined in runs, plenty of green cut clover should be given; also vegetables chopped fine, such as onions, beets, turnips, etc. Grit should be placed in trays and kept continually before them. In the line of grain, wheat, cracked corn and millet has proved a very successful feed. Generally, wheat should be fed in the morning and cracked corn at night, and when feeding three times millet is a very good feed for mid-day meal. Fresh water should be given them at least once a day during winter and twice a day during extremely warm weather. Earthen drinking vessels are best as they keep the water cool and are more easily cleaned. Great care must be taken to rinse and thoroughly cleanse the drinking vessels, as disease often comes from impure drinking water.

FEEDING YOUNG CHICKS.

BY EXPERIENCE, the best food for Bantam Chicks is wheat, fine cracked corn and oat meal. Youngsters should not be fed for the first forty-eight hours. The first feed should be dry bread crumbs or bread squeezed dry after being soaked in milk. This should be fed every two or three hours for the first day or two, then oat meal may be fed once or twice a day. After the first week a feed twice a day in place of bread may be given of Johnny cake made of two parts corn meal to one part wheat middlings, stirred quite dry with buttermilk or sour milk, adding a small teaspoonful soda to one quart of the milk. If milk cannot be obtained water can be used. This thoroughly baked makes a most excellent feed. When three to four weeks old feed fine cracked corn and wheat alternately once a day. At this age chicks may be weaned to three times a day, when the Johnny cake and the grain are the stand-bys. Fresh water should be given at every feeding, as chicks should not be allowed to drink water that has stood in the sun and become warm. Rinse and cleanse drinking dishes thoroughly. Charcoal and grit should be kept before the chicks from the first. Where chicks are not allowed free range green food must be furnished them. Never starve your chicks to keep them small. Size should be considered in the mating of the breeders before the youngsters arrive. By experience the above plan of feeding, watering, etc., has proved very satisfactory.

MATING BREEDING PENS.

PENS should consists of a male and four or five females. Yet some varieties, such as Japanese and Rose Combs, produce better results if bred in trios. Better results will be obtained by mating pens of this number than larger ones, as eggs are more liable to prove fertile and the chicks stronger. Inbreeding is practiced by nearly all breeders, but if new blood is introduced great care should be taken to purchase of reliable breeders and get birds that are properly bred.

HATCHING BANTAMS.

THE TIME for hatching Bantams depends largely upon what is required of the youngsters. When breeding for early fall shows or fairs we get the best results by hatching chicks the last of March or the first of April, as chicks for showing should be as near six months old as possible. When hatching for breeders and winter showing, chicks that hatched in May or June generally give good satisfaction. While these are the rules most practiced, yet Bantams may be hatched any time between April and September.

A WORD ABOUT LICE.

LICE HAS destroyed many a valuable bird. Bantam chicks are easily succumbed by this vermin. Inspect your birds often, both old and young, and keep them free from lice by the use of some good insect powder. Persian insect powder has been proven very successful in ending the existence of these tormenters.

PART I.

Game Bantams.

THE Game Bantam is the most popular of all varieties of Bantams. They afford great study and pleasure in breeding. The true standard bred game bantams are much admired by fanciers and breeders for their high station, their long, thin neck tapering from the body to the head, small whip tails and closely feathered body. Game Bantams afford great pleasure in breeding as they become very tame and great pets. They are very tough and hardy, enduring cold and severe weather much better than ornamental varieties. The last edition of American Standard recognizes nine varieties of Game Bantams, viz: Black Breasted Red, Brown Red, Golden and Silver Duckwing, Red Pyle, White, Black, Birchen and Malays.

Black Breasted Red Game Bantams.

THIS variety is the oldest and perhaps the most widely known of all Game Bantams. They are bred quite extensively throughout the country, yet but very few first-class specimens are to be seen. Many people breed them simply for pets and ignore the standard requirements. Therefore many different styled and shaped specimens may be found.

PEN BLACK BREASTED RED GAME BANTAMS.

When the Black Red Game Bantam is bred to the high standard of perfection a more beautiful bird is seldom seen. Their face bright red and the cock with well trimmed head and throat, beak dark horn, eyes red, head, neck, hackle and saddle of one shade of bright orange, pure in color and perfectly free from stripes or markings. Breast, body and stern a rich, lustrous black, shanks and feet willow colored, very smooth and free from all defects. The female in show condition should be what we call ground color, or golden brown penciled with grayish brown, the body color even and regular in marking, color golden brown shade penciled with dark brown or black, the breast a rich salmon shading into an ashy color beneath the legs. As a whole the color should be even and free from any dark or imperfect shading. With this combination of colors and marking in both male and female, it makes a beautiful pair for exhibition or showing.

Brown Breasted Red Game.

THIS variety is more rare than most varieties of Game Bantams. They prove quite hard to breed to get birds with correct markings. The face, comb, wattles and ears lobes are of dark purple. The cock bird should have lemon colored saddle and hackle. Other parts of the body should be black except the breast, which should be black and each feather evenly laced with lemon color. The color of legs and feet should be dark. The face of the female should be dark purple or black, beak black, head or neck golden or lemon color with a narrow dark stripe through the middle of the feathers. Back black, free from colored markings. Breast black, each feather evenly laced with lemon. Other portions of the body should be black. Shanks and feet dark willow or black.

PAIR BROWN BREASTED RED GAME BANTAMS.

Duckwing Bantams.

THESE are considered very beautiful by all lovers of game varieties of Bantams.   H. S. Babcock in the American Fancier, says of the Silver Duckwing as follows:

Among the Female Game Bantams, other things equal, I give the palm to the Silver Duckwing.   Or to put the matter in another form, the Silver Duckwing hen is the handsomest colored among the Game Bantams.   The soft silvery gray of the body, the silvery gray hackle with its black striping, the light salmon breast, these make a beautif ful combination, as restful to the eye as anything in animate nature. The gray body coloring is more or less distinctly penciled, adding to the delightful effect of the whole. But it is seldom that other things are equal. Silver Duckwings, are usually shorter in head, neck and limb, fuller in hackle and tail, than the Black Breasted Red, Red Pyle, Golden Duckwing or Brown Red, and the reason of this is not far too seek. The Black Breasted Red, being the most popular variety, has been bred with the greatest care, and skillful selection for many generation has succeeded in giving to it the desired characteristics. With the Black Breasted Red, the Red Pyle is often crossed to keep up the richness of the coloring in the males; with it the Golden Duckwing is often crossed for the same purpose, or Silver hens are bred to a Black Breasted Red cock to produce the Golden Duckwing; and even the Brown Red has probably received an occasional cross of the Black Breasted Red. In all of these varieties, the crossing with the Black Breasted Red has enabled them to gain some of the advantages that many generations of skillful breeding have bestowed upon that variety. In respect to shape I believe the Game Bantams should be arranged in order as follows: Black Breasted Red, Red Pyle Brown Red, Golden Duckwing, Silver Duckwing, White, Black, though it is possible that the Silver Duckwing and White varieties ought to change places, owing to the birds in the White variety that have been produced through the faded out Red Pyles.

But, despite the fact that the Silver Duckwing ranks below several varieties in shape, it by no means follows that it does in popularity. As a matter of fact, I believe the Silver Duckwing stands second in the list of varieties in popularity in this country, a position occupied by the Golden Duckwing abroad, I am led to believe. And it does this,

SILVER DUCKWING GAME BANTAM COCK.

probably, on account of its exquisite coloring. It certainly outranks the Golden Duckwing, the Brown Red, the White and the Black, and I think also the Red Pyle.

Although, in shape, I have placed it not high in the scale, it is not to be inferred that it is a little dumpy creature with no style about it.

On the contrary some specimens are produced with fairly long heads, with good station, good necks, real little games, attractive through their shape as well as their coloring. I have had hens that scored 95 points, and that, too, under good Game judges, and that is evidence enough to prove that fine shaped Silver Duckwings can be and are bred. But the 95 point birds of this variety are not plentiful.

Males that honestly score 91 to 93 and females 92 to 94 can be safely reckoned as first-class.

The males are not quite so gorgeous as the males of the Black Breasted Red variety, but they are quite handsome. The contrast between the black of the breast and the white of the hackle is sharp and clear, as is that of the silvery rose of the wing and the gleaming bar or the white of the back and the lustrous tail. The Silver Duckwing cock is a handsome bird, and the hen is still handsomer. If this variety equalled the Black Breasted Red in shape, there would be not a few to proclaim it the more beautiful variety.

The one thing, above all others, that Silver Duckwing breeders should aim for is shape. Select, select, select the very best specimens for breeders, make shape of prime importance, and I suspect it would not be a bad plan for such a breeder to mate up a very fine, light colored Black Breasted Red cock with some Silver hens and save the chickens therefrom, the males to breed back to Silver hens, the females to a Silver cock. This seems to me to be the quickest way in which to improve the shape. It is true that some defects in plumage would follow, rusty wings and the like, but these defects could be bread out and the substantial gain in shape preserved. At any rate, the experiment, I believe, is worth trying, for, improve the shape and increase of popularity is certain.

PAIR GOLDEN DUCKWING GAME BANTAMS.

Golden Duckwing.

THIS variety resembles the Silver in every respect except color, the cock bird having gold instead of silver on wings. Hackle and saddle being straw color. The female differs only in color of breast, which should be a deep salmon.

PEN RED PYLE GAME BANTAMS.

Red Pyle.

THESE birds, when properly bred, make very beautiful specimens. The male coloring is red and white. This combination of colors, with clean yellow legs, makes a very attractive bird. The female, when properly bred, should be pure white with golden hackle centered with white. The color of the breast salmon shading into white down the thighs. Where the feathers are white the clearer they are the better. In years past willow legs were bred in plenty, but as the new standard prefers yellow legs, the willow legged birds are fast becoming scarce.

TRIO WHITE GAME BANTAMS.

White Games.

MR. H. KOCHERSPERGER favors me with an article on this variety, and as he is an experienced breeder of the White Games, the article, which follows, will be of much interest:

The White Game Bantam is one of the varieties of Game Bantams which has been much neglected. Many people have been deceived by the introduction into the show-room of White Bantams shown as White Games, when in reality they were nothing but faded White Pyles and Brown Red sports. I contend that it is not right for a judge to award prizes to such exhibits as these. The existence of this circumstance is a disgrace to the breed.

What we want is enough breeders who will interest themselves in forming a club to try and make this breed more perfect, and to request county and state fairs, poultry associations, the National Bantam Association, the American Exhibition Game and Game Bantam Club to offer specials for this class of Game Bantams. In the past no class has been assigned to this variety, and this inconvenience has caused much annoyance to breeders and has discouraged many exhibitors. As they are a recognized breed in the American Standard of Perfection, why should they not have a separate classification and be placed on an equal footing with the other sorts? There is no good reason, and if the breeders will get together they will be able to gain for this beautiful Bantam adequate consideration.

Now I, for one, am willing to contribute $5.00 each year towards a fund of $25.00 to $50.00 to be awarded as specials by the National Bantam Association, divided in such a way that the first, second and third best American bred White Game Bantam Cockerels and pullets shall get the benefit. Such birds are to be pure white, good Game shape, with shanks as nearly yellow as can had, yellow beaks and red eyes. Birds not approaching the standard requirements are not to be placed.

Some such plan as I have endeavored to outline would surely cause the older breeders to renewed efforts and the new breeders would take a lively interest. We must admit that this breed is a difficult variety to perfect. To obtain the yellow shanks and retain a gamey shape are no simple tasks. But with all that the Game men have to contend against there is hope that they will, by their American energy and pluck, succeed in establishing this Bantam in the hearts of many lovers of a white bird. The inducements of specials for the best specimens will be a decided encouragement, and I have hopes that this breed will become more popular. It is to be hoped that the secretary will endeavor to obtain from the prominent poultry shows a separate class, and I sincerely trust that those who are interested in White Game Bantams will correspond with the secretary, stating what special they will be pleased to offer at the next meeting of the National Bantam Association.

H. KOCHERSPERGER.

Doylestown, Pa.

TRIO BLACK GAME BANTAMS.

Black Games.

THE Black Games originally came from Brown Reds. The plumage should be a deep lustrous black. The color of face a dark purple. Beak very dark and eyes black. This variety of Game Bantams is very rare and it is hoped more interest may be taken in their behalf as they can be made prominent and a most attractive variety.

TRIO BIRCHEN GAME BANTAMS

Birchen Games.

THIS variety resembles the Brown Red to a great extent, as the coloring is silvery white where the Brown Red has lemon color. The cock bird should have a dark purple face, hackle, saddle and back silvery white, breast black and each feather evenly laced with silvery white. The other parts of the body should be black, shanks and feet dark. The female has a dark purple face like the male. The body should be black, neck and hackle silvery white with a narrow dark stripe through the middle of the feathers. This beautiful variety has come to the front, being very popular, and may be seen in most show rooms.

Malay Games.

THIS variety, not being very well acquainted with, is the only variety I have omitted to illustrate. They are very rare and seldom seen. They are a little larger than the ordinary Game Bantam. They resemble the Black Breasted Red Game in color. They should have a lump comb, face, wattles and ear lobes red, beak yellow; also legs and feet yellow.

Bantams and Their Breeders.

Mr. Latham, Flatbush, Long Island, N. Y., Secretary of the National Bantam Association, favors me with an article in the interest of "Bantams and Their Breeders."

Your request that a few lines be written in the interests of "Bantams and Their Breeders" is received, and I take pleasure in assisting, if possible, the work of promoting the Bantam Fancy.

The more one looks at it the stronger inclined he becomes that Bantams are of so much importance as to warrant the amount of time and money which are annually spent for the improvement of these dainty fowls. They are deserving of much attention by careful breeders, for they will repay any fancier who will earnestly work for their advancement. The real pleasure derived from the labor of handling a Bantam plant is, in a large measure, an education in itself. The constant desire to bring everything to a nicety, in harmony with the Bantam Stock, has an important effect upon the owner of such a plant. All who have such surroundings can pride themselves, and those who are not so fortunate must content themselves by planning for improvements which, when the opportunity shall present itself, can be perfected. But with it all, the fine show birds, the excellent breeding stock, the fancier, if there be no interchange among the breeders, if they remain apart and are not inclined to associate with one another, the Bantam Fancy will remain, as it were, in a stagnant state. There can be no life, no energy.

If such a condition exist for any period the effect upon the fancy would be demoralizing. But happily, and for the good of all concerned, such a possibility is impossible, simply because the Specialty clubs are constantly endeavoring to advance the interests of Bantam Fanciers, and prompting them in every move that they make to stimulate the Bantam industry; and the indications now point toward a steady advancement of these Midgets, mighty in miniature.

The helpful effects of the National Bantam Association, which the writer has the honor to serve, are apparent in Bantam Centres, and, while we do not contend that this association is "the" one for Bantam men to join, yet we do claim that the National Bantam Association embraces every variety of Bantams and has a membership which is composed of many of our most promising fanciers in this country, and enjoys the confidence of the entire fancy. What we desire to have the Bantam men do is to consolidate and boom Bantams.

Join some organization which promotes the cause and, if you be favorably inclined, we shall be pleased to welcome all who desire to become affiliated with us in the progressive movement of developing the interests of Bantam culture in America.

E. LATHAM,
Secretary National Bantam Association.

PART II.

Ornamental Bantams.

THESE Bantams in most varieties are considered great pets. Many of the varieties are odd in appearance while others are noted as being very ornamental. Bantams in most cases are exact miniatures of the larger corresponding breeds of fowls, therefore, being small and easily handled makes them the most attractive of the feathered tribe. The present standard recognizes 16 varieties of Bantams other than Game, viz: Booted White, Cochin, Buff Black, Partridge and White, Brahma, Light and Dark, Golden Sebright, Japanese Black Tailed, Japanese Black, Japanese White, Rose-combed Black, Rose-combed White, Silver Sebright, Buff Laced, White Crested White Polish.

Housing Bantams.

WRITTEN for Turner's Bantam Book by L. L. Lucas, Oil City, Pa.

HOUSING BANTAMS—MY IDEA.

We all of us, as the Bantam readers well know, have our own ideas in regard to housing our own little pets. The only way to learn is to state our ideas, so I shall endeavor in this article to tell you mine in regard to erecting and building Bantam Coops. First of all, Bantams require just as nice and warm a coop as larger fowls. Some breeders think that a large dry goods box with a window and door in same will answer, but I find from experience, especially here in our climate, where the thermometer registers often 10 and 15 degrees below zero, that it requires a very warm and dry coop, especially where one desires to obtain the desired successful results. I build all my coops double boards, with tar paper between and double floors, with windows (not large ones) to the south. Too large windows heat up during the day and freeze at night, thereby making it much colder in the coop. Have by all means board floors made very tight and at least three inches from the ground. A cement floor is very cold and besides is liable to gather moisture and become damp. Dampness is the cause of 60 per cent. of the deaths of these little birds, so by all means guard against dampness. This is why I advocate double boarded coops with tar paper between, as single board coops will be cold and damp. They are all right in summer but will not do in winter. The largest coop I have at present is 9x20 feet, divided into four pens with a hallway 3 feet wide running the length of the building. This coop is built of hemlock lumber one-half surfaced, tar papered and ceiled inside with a two inch air space all around and cracks outside are stripped. It is as warm as can be, the water having not yet frozen therein. Large coops, where one has the room, are the most economical to build because you save the ends and the room on which they stand. A breeder who is handy with tools can erect a coop as above mentioned, 9x20 feet, at a cost of not over $40.00. This will house 50 Bantams comfortably and will last many years if kept painted and the roof tarred. I have besides this coop eight others of various sizes and shapes in which I raise and house the cutest little fowls on earth (the Buff Cochin Bantam).

L. L. LUCAS,

Oil City, Pa.

Care of Bantams.

BEST BIRDS ARE THE CHEAPEST. BUY ONLY THE BEST.

AT THIS period of the breeding season we are all interested in the welfare of our chicks, particularly their feed, care, etc. Allow me to say I do not pose as authority on this subject but will give some idea of the method practiced at my yards. I never feed the little chicks for the first 24 hours, this would mean nearly two days from the time they were released from the shell. Always leave them under the hen in nest until they are thoroughly nestled. After taken from nest thoroughly dust with some good insect powder and put in a clean coop. If coop has been previously used, be sure to have it thoroughly cleaned and disinfected before confining the chicks in it, and have the floor covered with good coarse sand, also have a low dish with a mixture of charcoal, bone and grit before them at all times. Their first feed is usually a few dry bread crumbs for the first day or two, after this moisten the bread with milk, do not have it sloppy but slightly moist, also give pin-head oatmeal. This I continue for the first week or two at intervals of every three hours, and only just what they will consume, and not have any left over to sour and get contaminated by filth. After the first week or as soon as they will eat it, feed cracked wheat and corn. L have the corn cracked very fine, all hulls and meal sifted out. As they grow can dispense with the cracked wheat and feed it whole. Last season I used the "Fidelity Food" and liked it very much, it is composed of various varieties of granulated grains, meat, bone, charcoal and grit. Give fresh water from the start three or four times a day in a dish so arranged that they can only get their beaks in it. Should the chicks become dumpish look for lice, which invariably is the chief cause of the ailment and kills more chicks than all other diseases combined. It is a good plan to dust the hen and chicks with some good insect powder every two or three weeks, and pour kerosene oil in the crevices of coops even if you do not notice lice. "Prevention excels cure." As soon as chicks are large enough to roost, see that the coops are provided with a perch. They will keep much cleaner and healthier than if left to huddle in bottom of coop. I would advocate that Bantams not be bred later in season than June or July and fed all they will eat while growing. They may be a trifle larger when matured than a stinted, half starved specimen, in my experience cannot say that they are. Certainly they are worth much more for breeding and less liable to disease than a late hatched, stinted, undeveloped chicken. Too much cannot be said about late hatched, starved Bantams in order to keep them down in size. They are not fit for the show room, much less for breeding. Understand me I like birds bred within the weight limit, think they can be and know they are. If new blood is not introduced too frequently I do not think you will be troubled with over-sized Bantams.

Never leave the chicks in the small coops after the nights become chilly in autumn. Confine in winter quarters and protect them by not letting them out in the cold winds to get chilled and contract colds. Of course they should be let out on pleasant days. When confined in winter they should have chaff or shavings on floor to scratch in and to protect their feet from the cold floor. Their hard feed should be scattered in this and let them work for it, which will give them exercise. Would not advise feeding soft food to Bantams, although the larger varieties will thrive on it. In my experience Bantams will do better on hard grains, such as wheat, cracked corn, hulled oats, with plenty of bone, charcoal and grit. Clean quarters with plenty of fresh air and water are the essentials to success with Bantam raising.

In breeding the different varieties of Bantams the same rules must be adhered to as in breeding the larger varieties, viz: not go to extremes in mating, i. e., for color mating. For instance, in mating the Buff varieties, would not advise mating any light or lemon colored females to a red or very dark male, from such a mating you will not get any specimens of a uniform color throughout. Should have your matings, both male and females, of one harmonious shade as near as possible, in order to produce the uniform colors in all parts. I do not wish to infer that all will be perfection from this last mating, but know from experience that the per cent. of good ones will be much larger. By all means do not breed from a red or mahogany colored male. Such birds are a detriment to any breeding yard and should not be tolerated at this age. If you have this sort better kill them and start in with the right kind, which can be obtained from many breeders; it will be cheaper in the end and the result more satisfactory. This extreme mating will apply to all varieties, the solid, as well as the parti-colored breeds. I would not advise breeding more than four females to a male. If you haven't this number of good females breed from less, had rather breed from one good pair than to breed from a dozen fair to medium. If you have no good ones, kill them and buy some, it will be far cheaper in the end. There are plenty of good birds in this country that can be purchased at a reasonable price. Never buy a poor bird simply because the price is low, the best ones are always the cheapest. Yours for the improvement of Bantams.

H. J. QUILHOT.

Johnstown, N. Y.

Success in Breeding Fancy Bantams.

WRITTEN FOR TURNER'S BANTAM BOOK BY L. L. LUCAS.

A MAN said to me some time ago, "Why don't you have those houses (referring to Bantam houses) made into conservatories and therein keep and grow flowers, they are more beautiful and you could make far more money?" I said, "My dear sir, it takes all kinds of people to make a world; you love flowers, fine, elegant, fancy flowers; I love those beautiful little living flowers that respond so readily to your kindnesses." By keeping only one kind a breeder, by devoting all his coops and time to the one breed, is by far more sure of success. Not only thorough care is required, but superior intelligence and always a good knowledge of the affair undertaken. This is far more true in breeding fancy fowls than in anything else. I would not advise anyone to undertake the breeding of fancy Bantams with a view to making much money, unless he has gotten a good insight into the laws which govern fowl reproduction. As a rule, a beginner enters into it with a great deal of zest, but nine times out of ten will fail only to try it again and again until experience has taught him what a difficult art it is. Still, he is bound to succeed in time if he has lots of grit and pluck, a level head, and above all, a natural love for his little feathered pets. Thoroughbred fowls and fancy fowls are not the same, as some seem to think. Fancy fowls means birds bred to standard requirements and as pure bred as it is possible to make them by keeping only the best and best blood in them. Fancy Bantams command high prices and eggs from high scoring prize winners command equally high prices. This is the secret of success as far as the money goes in Bantams, that is, by keeping only high class fancy birds. The real fact of the matter is that the growing of fancy Bantams is a difficult task and requires experience, study and patience. But this is largely offset by the profit there is in it when once mastered.

L. L. LUCAS.

Oil City, Pa.

Bantams at the Shows.

WRITTEN BY B. C. THORNTON.

I AM PLEASED to note the great increase of Bantam fanciers and exhibitors of them the last few years, and I must say that part of this is the result of the earnest and hard work of our very estimable secretary of the National Bantam Asciation, Mr. E. Latham, who has worked with great zeal in his endeavor to bring these little pets to the front, and that he has succeeded the great increase of entries in the Bantam varieties at our large shows, both fall and winter, will prove.

It has not been over eight or ten years since when an entry of 50 to 75 Bantams at our leading shows was considered a fine display, but look at the entries now at our large fall fairs and winter shows. The entries now number from 300 to 400 and are a grand display in themselves, and are one of the main attractions of the poultry exhibit.

And when one looks at the great improvement that has been made in all varieties of Bantams it is no wonder they have become so popular as they have. They are now bred the exact counterpart of their big brothers and sisters and are as fine in shape, color and exhibition points as the larger breeds.

The Cochins and Game, I think, have made the greatest stride to perfection as exhibition birds. I know that Game Bantams that were considered grand specimens no longer ago than four or five years would not have a ghost of a chance to be in the money now at any of the great fall fairs.

In speaking of fall fairs, I don't wish anyone to think that only second or third class birds are shown at them, because if they do they will be left sure, as the competition at the great fairs has become so strong the last few years that the best birds of the year are shown, and they must be if one expects to carry of the honors.

There has been quite a boom the last two or three years in Brown Red and Birchen Game Bantams, and they now deserve it, as there has been great improvements made in them in shape, style and color. The old coppery colored Brown Reds are a thing of the past; they are now bred to the fashionable lemon color, with handsome laced breasts and good body color.

The Birchens have also improved very much; instead of a creamy or yellow white they are now a pure silvery white and good black body color. I know of a pair of Birchen chicks that were imported in February of this year that Mr. Anderson received £20 for that are the best specimens of this variety I have ever seen; the cockerel is very fine in style, fine head, very good neck and wing, short body, very neat in tail, pure silvery white back, and the body color a superb intense black; the pullet is the equal of her mate in all ways, and they are a really beautiful variety when you see them—the correct color, and this pair certainly fills the bill. I would like to own them; they are toppers sure.

May the Bantam fanciers and exhibitors increase the next few years as they have done the last few and we will see an exhibit of seven to eight hundred specimens at our leading shows. So mote it be.

B. C. THORNTON.

Importing Bantams.

BY CHARLES T. CORNMAN.

I AM ASKED to contribute an article on my experience in importing Bantams, so here goes to the best of my ability. In the past fifteen years many of the best Bantams that have been imported into this country have found their way into my breeding yards by direct importation, and take them all in all as they pass before my mind's eye as a panorama of feathered beauties, I am content. Sure there have been disappointments, but in many instances they surpassed my expectations, so take it all in all I extend to my brother fanciers across the pond my hearty thanks. The only real disappointments I have experienced have been in Cochin Bantams, especially in Buffs and Blacks. I find that color does not count on the other side as it does here, and birds that are disqualified by our judges on account of white in plumage, if particularly good in other points will win under English judges. Fifteen years ago it was almost impossible to get strictly first-class specimens of Bantams at home, but not so now. Of late years I have seen many of England's winners displaced by American bred birds at our leading shows, and I trust in the near future it will be necessary to import but few specimens. Sure we have many complaints from parties importing birds to the agent, that they have been roasted by our English brethern, but in the majority of instances they have expected gold dollars for fifty cents each. You cannot invest ten shillings and expect ten dollars in return. The greatest drawback to importing birds has been the debilitating effect of the journey, on account of which a fair percentage of birds is lost of each importation. The direct cause of this is the careless manner in which transportion companies care for them during transit. Two years ago I imported some choice specimens in Game Bantams and my eagerness to see them took me to New York so I could be on hand immediately on arrival of the vessel. As soon as possible I went aboard, and after diligent search found the birds in the vessel's hold, dark as midnight, and impossible to see them without the use of a lantern. On inquiry I found they had been there ever since leaving Liverpool and had not seen a ray of light during their confinement of eight days, except when the steward held a lantern a few moments before them to give them a chance to hurriedly gulp down enough feed and water to keep life in them. Fortunately all my birds were alive, but several consignments that I noticed had dead birds in them, and in conversation afterwards with the consignees, I found that others had died after landing. Surely when you consider the excessive charges made by these companies better treatment should be given them. One reason for their careless and indifferent treatment is the knowledge that they are not open to any loss even though the whole consignment should die. I have found that birds arrived in much better condition when forwarded by fast freight steamers and a considerable less cost. Sure, it takes four or five days more, but what does that count when the health of the birds is considered. These ships are fully equipped for handling live stock of all kinds and as a consequence the birds get better treatment. See that the steward receives a liberal fee and you will find it a panacea for many ills and disappointments.

CHAS. T. CORNMAN,

Carlisle, Pa., May 7th, 1900.

TRIO BOOTED WHITE BANTAMS.

Booted White Bantams.

THIS variety is as the name implies, very heavily feathered on the shanks, having long vulture hocks or stiff feathers extending from the thighs. The longer these feathers and the heavier booted the more valuable is the bird. In color they are pure white. The head is medium sized having a single comb evenly serrated. Wings large, drooping slightly. Tail should be carried upright, shanks should be white. They seem tough and hardy and can be raised easily,

Cochin Bantams.

"COMMON DEFECTS," BY DR. WILLIAM Y. FOX.

THERE is one defect that is common to all varieties of Cochin Bantams and is also common to all individuals, at least to all males. That is, the head is too high in comparison with the cushion. This is partly due to too great a length of neck and partly to the proud way all Bantams have of carrying themselves. Thus the keel, breast and head are up while the tail is down. I have never seen a Bantam male free from this fault and few females. We have all seen pictures of Cochin Bantams that look exactly like large Cochins and can only be known as Bantams by comparing their size with that of some object in the picture. Do you not say at once, "that is an idealized picture, no Bantam ever yet had that carriage in such perfection?"

One other defect that all have is too large combs and wattles, frequently so large as to be really grotesque and rarely ever as small as they really should be. Stiff hock plumage, although not a disqualification, is pronounced objectionable by the Standard, and objectionable it certainly is. A Bantam male with really soft hock feathers I have never seen. To be sure we get quite a number so well furnished with loose fluff that the hocks can by no means be called vulture. In females I have seen several without the least stiffness in hock plumage, but the vast majority have the faults just as the males do. Let us now consider the Whites, Blacks and Buffs separately. In the Whites the worst and most common defect in color is a yellow tinge, and only a few years ago it was impossible to find either a male or a female free from this fault, while now at our large shows we see as pure white plumage on Cochin Bantams as on any white breed in the show. This has been accomplished, too, without sacrificing the bright yellow of shanks and beak. Many of these pure white males will, at the end of the breeding season have a brassy or sun burnt surface, which is entirely different from the yellow tinge which colors the whole feather and shaft clear to the skin. Another fault in color is the presence of foreign colors, either in part or the whole of the feather. Black, red, yellow and brown are all seen occasionally, but the commonest is black which usually appears in splashes and specks. It is a noticeable fact that a bird that shows a lot of these black ticks is almost always pure white without the slightest yellow tinge. On the other hand a very yellow bird will not show a black speck anywhere. Most of the Whites now have good yellow shanks and beaks, but five years ago white beaks and white, blue or green shanks were almost as common as yellow.

Their worst faults in shape are traceable to their descent from or at least mixture with the White Booted Bantams and are long, pinched tails, still very common, and long legs as well as very stiff hock and foot feathers. The Blacks have a very serious and very common defect in color, viz, white undercolor in males, birds fine in surface color, when opened show pure white undercolor in neck and sometimes in saddle. I believe this fault was imported from England, where undercolor cuts no ice, I am told. I bred Blacks five years before I ever saw a male black to the hide, and in all that time I never saw but one female, other than solid black and that one had white undercolor in breast. I believe Hon. Dave Nichols was the first to show cockerels black to the skin, and I think that was not more than five or six years

ago. To-day most of the cockerels exhibited are solid, but a
cock that shows no white is still quite a rare bird, some of
the very best colored cockerels moulting with considerable
white undercolor. Males also have a tendency to show straw
color and red in the surface color of neck and also sometimes
on the wingbows. This is not now a common fault, but the
objectionable purple cast in both males and females is very
prevalent even to-day. Some females also show a perfectly
dead black with no lustre and more of the desired greenish
shade. While this is not so bad as the purple, it is not right.
White in flights as well as foot feathers we find occasionally
of course. I have come to believe this defect is often due to
faulty nutrition in the growing chicks as much as to inheri-
tance. A flock of black chickens that struggled with lice,
poor food and poor quarters will show lots of white feathers,
while chicks out of the same parents and well reared will be
all right. I have also observed that the late hatched chickens do not come as true to color as their earlier brothers and
sisters. This may be partly due to the breeding stock not
being in so good a condition late in the season, but I believe
is mostly the result of the slow growth of the chicken in the
cold weather. White shanks are seen too often, in fact, too
little attention has been given to the color of shanks in the
Blacks. Some breeders to-day prefer yellow shanks to the
standard color, but black, shading into yellow, with bottoms
of feet yellow, seem to me the most natural color, and surely
as that is what the Standard calls for I cannot understand
anyone's reason for demanding yellow legs. In shape the
Blacks have no particular defect other than as mentioned in
defects common to all colors.

The defect in color of the Buffs is unevenness. What-
ever our understanding of the shade described as "rich gol-

den buff" we can all agree as to the meaning of "one even shade." In the males it has been almost impossible to get this. The neck hackle is one shade, the saddle hackle two or three shades darker, while the wing-bows and back are red, and the breast any shade it happens to come. Of course we do see some that are not quite so bad as this, but the average run of birds are fully as bad as described. Just why this very serious fault has been allowed to run so long without correction I cannot understand. The judges are partly to blame, for they have scored such birds too high altogether. I have had a cockerel showing four distinct shades of color, score 96 in a first-class show, and only cut 1½ points on color. Any breeder who did not think for himself would naturally say "that bird is all right, if I can keep on breeding them as good as that I have got the real thing," whereas he is not within four miles of standard color.

Some say we must keep the cherry wings or we will get white in flights. I do not believe it yet, and shall struggle for "one even shade of rich golden buff" a good many years before I will allow that white flights are a necessary accompaniment of this much desired end. In the females mottling and shafting are both common, as is also too dark a neck hackle. Black and white feathers are both seen occasionally in the Buffs of both sexes, but certainly are no more prevalent than in the large Buff Cochins. Green and blue shanks are seen too often especially in the pullets, and white shanks come in both sexes once in a while. It is a very discouraging thing to pick out the very best colored pullet in the flock and find that she has green legs. But those are most always the ones that have that defect. Can it be that green is the natural shank color to go with the buff plumage we all want? Perish the thought!

PAIR BUFF COCHIN BANTAMS.

Buff Cochin Bantams.

WRITTEN FOR TURNER'S BANTAM BOOK BY L. L. LUCAS.

COLOR.

As bred now-a-days, the Buff Pekin Bantan cock or cockerel is of a much darker shade than the hens and justly so, as from experience I find by breeding a dark male bird to a light hen will produce a greater number of solid color standard birds. A light colored male, called by some breeders cinnamon color, bred to females of his own color produces much better males than females, and as the percentage of males hatched as a rule is greater than that of females, such a mating proves very unprofitable. The color of the Buff Cochin Bantam, male or female, should be pure, whether light or dark buff. Let there not be a speck of white or black in any part of the plumage or in leg or toe feathering. These little Grains of Gold bred thus, will produce most satisfactory results.

L. L. LUCAS.

Oil City, Pa.

PAIR BLACK COCHIN BANTAMS.

Black Cochin Bantams.

THIS variety of the Cochin family is considered the hardest to breed. To get a bird that is black to the skin in this variety, with other points to correspond, is not an easy task. When you get a bird that is good Cochin shape, heavy toe feathers and good head points, many times you will have a gray or light undercolor, while to the reverse when good undercolor is to be had the other points are lacking. Therefore one must be very careful to start with strictly first-class breeding stock. When this is done and a nice flock of chicks raised the owner will realize a neat sum of money, as Black Cochins that are black are always in great demand and at first-class prices.

PAIR WHITE COCHIN BANTAMS.

White Cochin Bantams.

THESE are considered the toughest and hardiest of

the Cochin family and are easily bred. Many fanciers line-breed these for many years, which has proven very satisfactory. The color should be pure white, free from creamy shade. Cream color in White Cochins, like all other white birds, is one of the principal defects and should be carefully looked to in selecting the breeders. These, like the rest of the family, should be miniatures of their larger cousins.

TRIO PARTRIDGE COCHIN BANTAMS.

Partridge Cochin Bantams.

T. F. McGrew in the American Fancier, says:

THE almost endless task of perfecting a new breed or

variety of Bantams is but little understood by the majority
of fanciers.   We all know how few real good specimens come
during a season from the most carefully bred strains of the
best established breeds.   We presume there is not produced
in any one season in this whole country one hundred really
good specimens in Buff Cochin Bantams, this being true of
one of the very oldest varieties of Bantams.   What may be
hoped for in the establishing of an entire new breed?

Mr. Entwisle of England was the original producer of
Partridge Cochin Bantams.   Others have started their production but never completed the task.   Mr. Entwisle united
the Buffs and Blacks, and crossed the result with large Partridge Cochins.   The color of the Partridge Cochin of England is so foreign to that demanded with us, that the color of
the females that come from them would be called badly off-
colored with us.   They are much the same color as some of
the reddish colored Dark Brahma hens sometimes seen with
us.   We demand the reddish or mahogany color for Partridge Cochin females, while with them a much lighter shade
is recognized.

Five years ago we secured from Mr. Entwisle a trio
of Partridge Cochin Bantams, from these we produced a
dozen chicks, part of which lived to form the start of our
present strain of Partridge Cochin Bantams, the imported
male and one of the females not living into the second season.   Those produced from the trio we continued to breed
together, and with them one of our American bred Standard
females, and in this way we have established a strain of Bantams that now produce fairly good color as graded with our
Standard demand.   Each season color conditions improve,
but no one can hope to equal the rich mahogany color of our

Standard Partridge Cochins in so short a time.

For two seasons we found that both the Black and the Buff of original days would crop out, some of the chicks come off a stately blue color, while others showed some mixed color, as if a cross or mixed breed. All of this pointed to the upheaval of the union of two bloods. For two seasons past all this trouble has disappeared, and they now breed as true as any Cochin Bantams, but for shape the Partridge Cochin Bantams seem to excel all the other varieties as a whole, they all have good Cochin shape with proper cushion and fluff, also good leg and toe feather. The color of the male is rich and clear, the penciling of hackle and saddle very good, in fact some of them fully the equal of the larger breed.

None of our Cochin Bantams seem to be as hardy as they, cold and exposure does not tell so much on them as with other varieties, and while all Cochin Bantams are exceedingly hardy, these seem to stand the most severe weather rather better than others. They are most prolific layers and their eggs hatch well. We presume this superior precosity comes from the infusion of new blood while producing them. Their rich color and markings are most attractive and are quite suitable to all surroundings. They look bright and clear under all conditions, and can be kept quite presentable even within the confines of a smoky city.

All who know anything of poultry know of the color and markings of our beautiful Partridge Cochin fowls, the rich glossy red and black of the male so much like the Black Red Games, and the beautiful rich mahogany ground color of the female so richly penciled with darker color, places them among the most attractive of all fowls as to plumage. The same beauty of plumage is found with the Partridge Cochin Bantam, if anything the miniature form adds to their real beauty and lends additional attractiveness. But few good specimens of the Partridge Cochin Bantams have been seen with us outside of the Boston and New York shows. Some are shown under this title that do more injury than a little to the variety. We must remember that the same Standard description rules with them as with the larger Cochins. Clay colored or unpenciled breasts lost recognition many years ago, simple bars across the feather are not Partridge Cochin pencilings, and a mixture of color on cushion does not represent the proper penciling of our Standard that calls for penciling that follows the shape of the feather. All this goes into the make-up of the real genuine little Partridge Cochin.

There is another point often lost sight of, and that is that a Cochin Bantam should not have legs under them like stilts, long stiffy legs belong to what is called Booted Bantams. Some of our Cochin Bantams have a mixture of the Booted Bantams in them. This gives the long stilty legs that are entirely out of place on all Cochin Bantams, at the same time a Cochin Bantam should not be so short in legs as to look as if it did not have any legs under it at all, they should not look like creepers, nor should they stand too high on their legs, but just the proper Cochin leg in proportion to their size. There is a disposition on the part of some to pay by far too much tribute to the size of our Bantams, just so they are small seems to carry the greatest weight with some. If fully the equal of others in Cochin qualities then the smaller specimens should have the preference, if not the favor should go to the best Cochin in shape, for Cochin shape first, if within the limit of weight, should have the consideration. Another point to be considered is the matter of feather, close feathering is not proper with Cochin Bantams, long, loose cushion and fluff belong to them and should have considerable consideration when present.

PAIR LIGHT BRAHMA BANTAMS.

Light Brahma Bantams.

WRITTEN BY GEORGE W. HILLSON, AMENIA, N. Y.

THIS grand new variety, which has been originated but recently and represents the monarchs of the poultry yard in a miniature fashion, are bound to become popular as they become better known. For what is prettier than a Light

Brahma, with its gorgeous plumage, haughty airs and stately carriage, and when reproduced in a miniature Bantam that weighs 26 to 30 ounces cannot be excelled for beauty, style or character. As layers they are unsurpassed by any variety of the Bantam family, and will lay continually from December until September if they are given clean and roomy quarters. For the city fancier, whose quarters are limited, no better variety can be found, for they stand confinement well and can be restricted by a fence four or five feet in height. In breeding Light Brahma Bantams select a cock or cockerel with a nice pea comb, good laced hackle, tail, and true Brahma shape, with plenty of leg and toe feathering. To mate with him select two or three hens or pullets with as white backs and undercolor as one can get; also two or three hens or pullets with strong hackle, wings and tails, with dark undercolor, no matter if they do show considerable black in back and fluff. From the former birds you will get good colored pullets and from the latter good colored cockerels. The chicks come black and white from this latter mating, but will be O.K. when matured. The chicks are quite strong when first hatched and are not very hard to raise if handled properly. I find a hard boiled egg chopped up fine with bread or cracker crumbs to be the best feed for Bantams for the first two or three weeks. After the chicks are three weeks old, millet seed, fine cracked corn, or wheat can be given with good results. Keep free from lice and do not let the Bantam chicks out until the dew is off the grass and you will meet with success and derive much enjoyment from rearing a few Bantam chicks.

PAIR DARK BRAHMA BANTAMS.

Dark Brahma Bantams.

By PERMISSION this article was taken from the "Bantam Number" of the American Fancier, written by T. F. McGrew, New York City.

The latest production in Bantams is the Dark Brahma Bantam, the origin of which came to us from England in a very crude condition. Those first seen looked as if produced by a cross of a large Dark Brahma and Partridge Cochin Bantam. Many of the chicks produced from them showed considerable of the dull red shading in both males and females, in size rather large, in shape too short on legs and fashioned rather after our Cochin Bantams. In addition to

this they had vulture hocks, a bad fault found in all English
bred Brahma Bantams, both Light and Dark, a condition to
be looked for as they allow these rough hocks on all Brahmas in England. Within a year I have seen the best Dark
Brahma Bantams ever imported from England to this
country. They weighed from two and one-half to
three pounds each, of good shape and color, bad in hocks,
and poor in tail formation. They show considerable improvement over former importations, and it is quite reasonable to suppose they have better than those sent to this
country. We think there are but two lots of Dark Braham
Bantams in America that can be depended upon to improve
into meritorious specimens. One lot is being bred direct
from imported stock, the others are being bred in three lines,
one in direct line from imported stock, one a cross bred male
one-half English Bantam and one-half blood of a fine
Standard bred female, the other an imported male
with one-half bred females, the result of first cross
of Bantam male and the Standard bred hen. The
penciling and color of the half-bred females is beautiful, and so far the cross is free from vulture hocks,
in size about the same as seen in our Light Braham
Bantams, but few of them under Standard weight. The
head points and comb on many of them are fully equal to our
Standard bred Dark Brahmas. Considerable trouble will be
experienced in getting proper hock and tail formation, this
comes from the cross of Booted Bantams and Cochin Bantams in their early formation, which gives the narrow upright tail and extended hocks, which was not improved by
the use of rough feathered Brahams in England. All these
difficulties will be removed by crossing as above with our
Standard Dark Brahams. I shall not be surprised to see
some very good ones shown the coming winter, those at work

on them have made great advancement with them within two years, and now have good breeding stock to work with. More attention should be paid to Brahma type in both the Light and Dark Brahma Bantams. Cochin shape is not proper, the Brahma Bantam should be Brahma shaped, not low set and short on legs like Cochin Bantams. The close built Cochin Bantam with Brahma coloring is not proper, they should be true Brahmas in shape. While vulture hocks are not a disqualification for the present, they will be before many years, and should be guarded against. We know that some who handle them favor the low set, compact built type. I have heard some say, we prefer those with short legs and heavy feathers, but according to the reading of the Standard this is not correct and should be discouraged.

The disqualifying weights are rather severe on Brahma Bantams. A Brahma cockerel at Standard weight should weight ten pounds or 160 ounces, one-fifth of this is 32 ounces. The Brahma Bantam cockerel is disqualified at 30 ounces, to be at Standard weight on 26 ounces. It is admitted that one-fifth the size or weight of the larger fowls should suffice for the Bantam, but for these we disqualify if not under this rule as to weight. The only possible chance to attain these weights will be by very close breeding and under feeding, two methods that will destroy their vigor and vitality. We hope to see these matters adjusted and the disqualifying weights advanced a few ounces for the benefit of the Bantams themselves. It is for the best interest of all new breeds to favor them somewhat along the lines most difficult to obtain, for by so doing it makes it possible for efforts to culminate into success. If full quality is demanded for the new that is obtained in the old and well established breeds, it presents but little encouragement as you advance. So strong a rule demands perfection before any consideration, this detracts very much from the interest and retards development.

PAIR WHITE JAPANESE BANTAMS.

Japanese Bantams.

WRITTEN for Turner's Bantam Book by F. S. Bullington, Richmond, Va.

There are three varieties of Japanese Bantams, viz: White, Black and Blacked Tailed. To breed good Japanese

requires considerable amount of skill, and if a breeder gets
30 per cent of good colored and well marked chicks, he will
be well repaid and successful. This variety of Bantams are
the most peculiar looking of all, having very short legs and
with wings drooping nearly to the ground. The tail is very
large and upright and in good specimens the head and tail
meet. The Whites are pure white in color.

TRIO BLACK JAPANESE BANTAMS.
The Blacks are of a green, lustrous black throughout.

PAIR BLACK TAILED JAPANESE BANTAMS.

The Blacked Tailed are white excepting tail and wings. When wings are folded they show all white but when spread primaries and secondaries are dark slate or black, edged with white. Color of feet and legs should be a rich yellow, and free from feathers. This variety is so hard to breed that good specimens command very high prices

PAIR BLACK ROSE COMB BANTAMS.

Rose Comb Black Bantams.

WRITTEN BY RICHARD OKE, LONDON, CANADA.

THIS is probably one of the very oldest and most popular varieties of the Bantam tribe, and I believe it can truthfully be said there are none that surpass them for beauty, and although they are generally looked upon as a purely fancy variety, in my experience I have found them most excellent layers, and considering the amount of food consumed will **give returns** which will surpass many of the larger varieties.

They may also be classed as a non-sitting variety; true some old specimens will evince a desire to take to the nest, but very rarely. In these rare exceptions I have found them most excellent sitters and good mothers, but my experience has taught me that it is unprofitable to rear Bantam chicks with Bantam hens, for, as a general rule, they are too fussy and do not supply the amount of warmth required. I find common hens, from three to five pounds' weight, to answer the purpose very well, especially in this country where, it is often the case, we have damp days and cool nights. With ordinary precaution the chicks are easily reared and feather quickly, but during the attainment of their first feathers they should have good care and be kept free from vermin, or many will droop and die. I have heard my esteemed friend, "Billy" McNeil, when addressing some utility crank, remark "that any fool can raise big chickens, but that it took brains to raise Bantams," and there is a power of truth in the same statement, and I venture to say that a successful Bantam breeder can put Mr. Utility out of the business as far as the breeding and rearing of chickens is concerned, for to be successful with Bantams experience has taught him to be very attentive and employ all the very best wisdom known to the science.

I do not know that any set of rules can be laid down as to food, as it will vary under different conditions. It has truly been said what was one man's meat was another man's poison, and this same rule will apply in rearing Bantams. I have found if they are fed wholly on cooked foods for the first few days they are less liable to bowel troubles, which should not be overlooked, as ailments of this kind are very weakening to Bantam chicks. Anyone having a good method of feeding that they employ had better cling to it, as your own observations have taught you which answers your purpose best.

It is needless to say that Black Rose Combs have improved wonderfully in this country within the past few years. One has but to attend one or two of our leading winter shows and be convinced and admire the beautiful specimens now exhibited with an abundant length of feather, rich sheen of plumage, and superb head points, lobes, etc., combined with symmetrical carriage.

TRIO WHITE ROSE COMB BANTAMS.

Rose Comb White Bantams.

WRITTEN FOR TURNER'S BANTAM BOOK BY "ZIM."

THE Rose Comb White, and Rose Comb Black Bantams, when well bred, are among the very handsomest of the Bantam family. Until of late, say three to five years ago, the white variety was not popular, and no really good specimens were seen, hence one of the reasons why they were not popular with true fanciers, as it is an undeniable fact that the genuine fancier must have something beautiful before his admiration can be aroused, or he can be made to take an interest in it. There are some magnificent specimens of Rose Comb White Bantams in America to-day, and in consequence there is a great demand for them. Now, at our eastern shows at least, we see them of small size, with the jaunty style and carriage of the best Blacks, with fine combs and white lobes, and all the really good ones have white legs, showing pink through the scales. Our Standard allows shanks "white or yellow," but "yellow" should have been stricken out, as we do not believe anyone ever saw a really good one with yellow legs. All such that we ever saw were inclined to have a creamy tinge to plumage, and often a yellow earlobe, which was, or is, more liable to be ticked or edged with red, than those with white legs and skin. The White Rose Comb of to-day, with his proud carriage, pure white plumage, with its silvery gloss, large, smooth, pure white lobes, white shanks and diminutive size, has created scores of admirers for them, and they are here to stay, and the coarse fellows of a few years ago, with their ugly, irregular combs, long legs, pinched tails, long backs, yellow plumage, yellow lobes, full of red ticks, need expect no ribbons in competition with him, if handled by an up-to-date judge. The Blacks for some time past have been the dandies and belles of the Bantam alleys at our best shows, and in score card shows, if there was a special offered for the highest scoring Bantam, it was a fight, nearly every time, between a R. C. Black and a Black Red Game Bant, with odds in favor of the Rose Comb, and often, he won, if the special was for the highest scoring bird in the show, large or small. Within the last five years, even this breed has made advancements. They are bred smaller than formerly, with far richer and more perfect colored plumage. We see specimens now with rich metallic black plumage, entirely free from any bronzy tinge, or purple bars to the feathers, and but few are now seen at our best shows, or in our best breeders' yards, of that dull, lustorless, brownish black color. We could mention scientific men who are breeding these varieties, that have birds whose combs and lobes, as well as style, size and color, are almost perfect and can see one particular bird in our mind's eye which not only won in his class, but won special for best ornamental Bantam at New York, that we consider the nearest a perfect specimen of any bird of any breed we ever saw.

Bantam Briefs--Sebrights.

PAIR SILVER SEBRIGHT BANTAMS.

BY F. B. ZIMMER, GLOVERSVILLE, N. Y.

OF THE two varieties it would be hard to tell which is the more popular or the more beautiful. A well-bred, correctly marked Sebright, be it Golden or Silver, is certainly a very beautiful bird. But judged from the standpoint of an expert and critic, there are so very few produced that are extra good, all round, that not many specimens are met with each year that hold him spell bound and cause him to inquire for the owner and breeder, and figure on how much money he could pay for the specimen and have enough money left to insure him against walking home, as it is very hard for the man who is really capable of taking in all the good points of a real crack-a-jack to go home without it; and to such a man such a bird is cheap at any figure. There are coveted qualities in the Sebright that our Standard does not explain or even mention, that a few men we know of, among them men who have assisted very materially in bringing the Sebrights up to the high average of today, prize very highly. One of these desirable qualities is the shape of the feathers on breast, shoulder and back. The true Sebright feather, the sort that shows off the color and lacing to the best advantage, and consequently is the handsomest, is a rather broad feather with a round end. Not over seven or eight years ago many specimens were bred and shown with feathers of these sections long and narrow, and with ends of feathers nearly straight across, as though cut off square with a shears; occassionally one sees a specimen of this sort yet, but not often. Another very desirable quality is evenness of lacing. Because a bird has feathers on all sections, laced clear around each feather, does not make the specimen perfect in color, as the lacing may be uneven, i. e., narrow down the side of the feather and one-half or one-third wider at the ends, or it may be too wide all around the feather, or again the black lacing may not be intense enough, but a faded black or dirty brownish color. Again, a bird might possibly be perfect in lacing, and in shape anything but a Sebright, and could then easily be defeated by one close to perfection in shape with good average lacing. In shape the perfect Sebright should be moderately short in leg, very short in back, rather large in wing, which should be carried rather low, tail of good size, carried erect and spread like a fan, carriage stylish and saucy. One of the very hardest things to obtain in breeding Sebrights is to get good style and general lacing and at the

TRIO GOLDEN SEBRIGHT BANTAMS.

same time a clear tail. We have seen many specimens with clear tails, but many of them were not laced all around the feathers of the breast. However, there has been produced a few, and comparatively a very few, with clear tails and good general lacing throughout. In breeding Sebrights all these things must be kept in mind and judgment used in mating. You can wager a man has good, sound judgment as regards how to mate and breed poultry if he makes advancements in or even holds his ground in breeding Sebrights. A word of caution: If you ever, by judicious breeding, get a line of birds that produce a good per cent. of chicks that are good in style and lacing, do not put in birds of unknown breed when in need of fresh blood. Get the same blood lines if obtainable, and females are preferable to males to supply the new blood.

The real judge of poultry must necessarily be a man strong enough in will to throw all prejudices to the wind. No matter where the stock came from, who owns it or how good any one particular bird may be in any one particular section. The judging of Sebrights is no exception. The judge must have or "ride no hobbies" in the show room, no matter what he thinks or does in his selection or breeding yards at home. Because a bird may have an absolutely clear tail and because this quality is so hard to get, does not necessarily entitle this bird to the first prize, regardless of his other defects. Because a man may be a good fellow or a first-class guesser does not necessarily make him an impartial and capable judge. To make an expert judge and do justice to any breed a man must have studied the breed, not the Standard of the breed only, and in no other breeds will this be made more apparent to the expert breeder and the critic than in the judging of Sebright Bantams and others of our laced and penciled breeds.

PEN OF WHITE POLISH BANTAMS.

Polish Bantams.

WRITTEN FOR TURNER BANTAM BOOK BY "ZIM."

ONE of the most ornamental of the entire Bantam family, and at the same time one of the most difficult to breed to a high state of perfection, are the Polish Bantams. That a well bred specimen, with its correct jaunty style and large round crest is beautiful none will gainsay. They are not extensively bred and the old style, non-bearded variety are far the most plentiful of any of the varieties. By "old style" we mean those with single combs and white or flesh colored legs, as admitted by the A. P. A. to the Standard nearly or quite 20 years ago. After the writer of these notes had originated a strain of White Bearded Polish Bantams, that were in reality miniature Polish with all the Polish characteristics, including the prominent nostril, the V comb and blue legs, and had them admitted to the Standard, many fanciers could see but little beauty in the "old style" non-bearded variety, not so much for the reason that they were beardless, as that they were not correct in combs and color of legs. Why they were admitted in the start without these true Polish characteristics I cannot comprehend, for at that time they would have been easy to obtain. Of this we are certain, as very soon after they were admitted we obtained our first trio of them, and within two years from purchase of our first trio, we purchased the originator's entire stock and have bred them ever since, and they often bred chicks with blue legs, which we had to discard as the Standard called for white, and occasionally there came a chick with a V comb.

As we said before, after the advent of the Bearded White Polish Bantam with a V comb and blue legs it was thought best by some to drop from the Standard the non-bearded variety, and this was proposed when the committee to revise the Standard met at Fishers Island, and I rather think it would have been done but for the plea of the writer, who after giving reasons why it should not be done, asked for the same length of time (5 years) to change the style of comb and color of legs of the breed, as was given Cochin Bantam breeders to do away with the 5th toe and willow legs of that breed, and which request was generously granted by the entire committee. Therefore in 1903 all White Polish Bantams must have V combs and blue legs or suffer disqualification. We take this way of reminding the public, as no

PAIR BEARDED WHITE POLISH BANTAMS.

longer ago than at the last New York show, in conversation
with a breeder or exhibitor of non-bearded Polish Bantams,
we learned that he knew nothing about this matter, notwith-
standing the fact that he owns a Standard and breeds and
shows the variety. It behooves lovers and breeders of the
non-beards to "be up and stirring" if they want to be in
"the swim" in 1903 with this variety. We know of one
small flock that can qualify now, but even after this warning
we dare assert there will be plenty shown at the expiration
of the time limit that will be disqualified. Many fanciers
claim the Polish Bantams are hard to raise. We claim they
are not hard to raise, that they are just as hardy as the larger
breeds of Polish, but we do admit they cannot be raised in

large numbers with other varieties of chicks, as they are a crested breed and cannot see as well as other breeds, in consequence are run over and hurt by the other chicks, but if their coops are set off to one side, away from other chicks, or if given a yard by themselves, they will, as large a per cent of them live to mature as any variety of Bantams. They must be kept separate for another reason, they are slow feeders, particularly the bearded variety, from the fact that beard and crest bother them to an extent in seeing their food, and if with other varieties they do not get sufficient before the others have it consumed. Again on account of their crest they cannot prune their plumage as well as a crestless breed, and consequently must be watched and not allowed to become lousy, and when the cold fall rains come they must be kept sheltered. All these precautions and extra care must be bestowed on any of the large varieties of Polish. Therefore anyone who can raise any of the crested breeds can raise Polish Bantams. The true fancier, the interested breeder, does not call this trouble, but simply terms it proper care, and those not enthusiastic enough to see that their birds have proper care, had best not start or invest in the business.

Diseases of Bantams.

PREVENTIVES AND CURES.

BY PERMISSION OF DR. W. Y. FOX, TAUNTON, MASS.

Cold.

A common cold is probably the most prevalent disease the human family is subject to, and the same is true of Bantams. The first symptom is sneezing, then a discharge of clear, watery fluid from the nostrils and eyes; later, a slight loss of appetite and general dumpishness.

In itself a cold is of little consequence, but, as it is often the forerunner of roup, it must not be neglected. Cold is generally caused by drafts blowing across the roosts at night, or by filthy quarters. It may also be caused by dampness in the house or runs, or too much exposure to bad weather. Bantams can be allowed in their yards in very cold weather if the ground is free from snow and mud, but they are much better of in the house if there is mud or snow on the ground, or if it is stormy. In this respect they certainly require more care than the large varieties. The prevention of colds lies in keeping the flock in clean, tight, dry quarters.

The treatment is very simple. If only one or two are affected remove them from the rest and place in a coop where they will be warm and free from drafts. Get some camphorated oil, at any drug store, and with a small glass syringe inject it into the nostrils twice a day. This will generally effect a cure within a few days. If many of the flock are afflicted in this way it will be impracticable to treat separately and the first thing to do is to find and remove the cause of the illness. Having done this, keep a small piece of gum camphor in the drinking water and watch carefully for further symptoms. Do not allow the nostrils to become plugged by a crust, as they often will, because the discharge will be held back and act as poison.

After the nostrils have been obstructed a day or two the head will begin to swell and before we know it we have a case of roup to deal with. The injection of camphorated oil as already directed will usually keep the nostrils free and open.

Roup.

This is a contagious disease and generally begins as a simple cold. It is often fatal, and is much to be dreaded as it will sometimes go through the whole flock before the owner is aware that there is any serious trouble. It is difficult to say just when a cold turns into roup, but when the discharge from the nostrils and eyes becomes thick and sticky, and of an offensive odor, you may be sure that you have a case of roup. The next symptom is swelling of the head and eyes; frequently the eyelids will stick together, and if washed apart a large amount of fetid matter will escape. As these symptoms increase the bird is growing sicker all the time, more dumpish and has little or no appetite.

Roup may be prevented by good care and prompt treatment of every cold, but above all by care in introducing new birds into the flock. Whenever you buy a new hen keep her in quarantine at least two weeks, until you are sure she is in perfect health, before exposing your stock to the danger of contaigon. Bantams of a strong, vigorous constitution, properly housed and fed, will never have roup unless they catch it from some diseased fowl carelessly introduced into their house.

Probably the most common way for the disease to be transmitted from one to another is through the drinking water. Be careful to thoroughly clean and scald any drinking vessel that has been used by any sick Bantam, before using it again. It is doubtful whether the disease can be carried in the air, but give the well birds the benefit of the doubt and confine diseased ones in separate houses or rooms. It is unwise to keep an invalid in a room with a fire, unless you are prepared to keep him there until warm weather, for it will never be safe to return the convalescent patient to the unheated house after he has had the luxury of a fire.

The treatment of roup is, in the main, very unsatisfactory, although if begun soon enough it may save a valuable specimen. Keep the nostrils, eyes and throat as clean as possible. Get a bottle of listerine at any drug store, and put a tablespoonful into a glass of warm water. Inject into the nostrils, swab the throat and wash the head and eyes with it two or three times a day for the first four or five days. Feed with soft cooked food and milk.

If this treatment makes no improvement in the patient, kill him and burn his carcass. This is the kindest and best advice that can be given, for, although he may recover after weeks of dosing and pampering, he will still be a weak bird and the slightest exposure will start a discharge from the nostrils, which may contain the germs of roup and be sufficient to cause the disease in the flock to which he belongs.

A Bantam that has once had a genuine severe attack of roup is never fit to breed from, as his offspring will be sickly, puny chicks nine times out of ten.   If you are unwilling to take this advice, as you probably will be until you have tried to cure roup yourself, the next best thing to do is to continue to keep head and nostrils as clean as possible, stop the aconite and give one grain of sulphate of quinine three times a day and all the milk and whiskey you can pour down, every three or four hours.   By this time your pet will not eat and his strength must be kept up by forcing the whisky and milk.   Should your efforts prove successful and the bird begins to mend, leave off the whiskey and quinine very gradually and put enough tincture of chloride of iron into the drinking water to give a decided brown color; feed good cooked food and a little meat once a day.

Canker or Diphtheretic Roup.

THIS is a frequent accompaniment of ordinary roup, and is probably a different manifestation of the same disease. It is highly contagious to other fowls and possibly to man. Cases are reported where children have probably contracted diphtheria from fowls sick with canker, and also where poultry that have access to discharges from diphtheria patients have sickness with canker.   The one distinguishing symptom of canker is the appearance in the mouth or throat of a white or yellowish white cheesy membrane.   This may appear during the course of ordinary roup, or may come on suddenly in an apparently healthy fowl.   At the first onset one or more white spots, about the size of a pin head, may be seen either on the roof of the mouth or under the tongue, or, quite often around the opening to the windpipe. These spots grow very rapidly until, often times, the whole mouth is filled with a membrane that is usually glistening white, sometimes yellowish. When torn off it leaves a bleeding surface beneath. It is of very offensive odor. If this membrane extends into the windpipe the patient will soon die of suffocation. This is a disease that cannot be mistaken, as the appearance of the membrane is very characteristic.

The remarks on the cause and prevention of roup apply especially to canker and need not be repeated. The general treatment is also the same, but the local treatment is different. Instead of washing out nostrils and mouth attempts must be made to remove the membrane. This is often done by scraping with a piece of pine wood whittled to a convenient shape. After removing all that can be removed, without excessive bleeding, the parts should be powdered over with alum. A better way is to apply peroxide of hydrogen in full strength directly to the membrane, which will soon be eaten away with much less bleeding than in the other proceeding. After using the peroxide a few minutes, apply tincture of the chloride of iron in full strength. The mouth can be pretty well cleaned by either method, but the membrane soon returns and the process must be repeated often. When the membrane is in the windpipe it has to be left to nature, and almost always proves fatal.

Cholera.

At the present day this is an extremely rare disease in the United States. It is the most contagious of the diseases of poultry, generally killing the whole flock when it once gets a foothold. It is always caused by contact with a previous case, never originating in a yard without such contact or exposure.

The symptoms are excessive diarrhœa, first of a black substance as thick as tar, later by a thin, watery fluid which smells putrid. There is very rapid emaciation and prostration, death frequently occurring within thirty-six hours after the commencement of the disease. There is no treatment; kill and cremate.

Diarrhœa.

This is frequent, and it is sometimes mistaken for cholera, but cholera is so very rapid that this mistake ought not to be made. Diarrhœa is usually caused by improper food, impure water, sudden changes in temperature or exposure to cold and wet. Individual mild cases require no treatment, as they will soon recover. In severe cases, remove the patient to a coop, keep without food for twenty-four hours, keep lime water before it instead of clear water. After twenty-four hours give a little bread, soaked in boiled milk. Let this be the only food until diarrhœa ceases. When there are a number of cases in the flock, be sure there is something wrong in food or drink. Search carefully for this cause and remove it.

Crop Bound.

This is quite common in Bantams, and if not properly treated is very apt to prove fatal. The first symptom is a constant effort to swallow. The neck is stretched out, the mouth opened, and the hen acts the way you often see a little chick act when trying to get down a worm one size larger than his gullet.

If you suspect that you have a case of crop bound, place the subject where he cannot eat for twenty-four hours and then feel his crop; if it is hard, or harder than when he was shut up, your suspicions are confirmed.

This trouble is caused by the plugging up of the outlet of the crop with some particle of food, such as a long, ribbon-like piece of hay or grass.

The treatment: Empty the crop. This can sometimes be done by pouring castor oil down the throat and working the mass in the crop around with the fingers. Try this about three times, two or three hours apart. If by that time the mass is not softened it is time to resort to surgery. Remove the feathers from a space the size of a silver dollar directly over the crop. With a clean, sharp knife make a cut one and one-half inches long through the skin; pull the wound along about half an inch and with a second cut go directly through into the crop. With a spoon handle scoop out the contents thoroughly. Either see or feel the outlet of the crop, so as to remove any obstruction there may be there. Wash the inside of the crop and the wound with warm water, to which a little salt has been added. With a needleful of white silk sew up the crop and then the skin. Give no food or drink for thirty-six hours, then give a little bread soaked in milk. Feed carefully for a week; by that time the little fellow will be all right, that is, supposing the relief to have been given soon enough. For, if the mass in the crop had fermented badly, as it will in three or four days, it will have excited so much inflammation that the operation does no good. Do not delay in a case of crop bound as twenty-four hours frequently make the difference between saving and losing a valuable bird.

Leg Weakness.

THIS is most common in growing chickens and is shown by inability to stand up. The chicken appears hungry, and all right in every way, except that it tries to get around on its hock joints instead of its feet. This occurs either while the first feathers or the second are growing. It is due to defective nutrition and is analogous to what we frequently term in children as growing too fast for their strength. The remedy is to change the diet, give more meat and cut bone, something to make more muscle. Take care that the other chicks do not prevent the weak one from getting any food at all. With a little care these cases recover in a few days.

Scaly Legs.

THIS is a most digusting affection and its presence in a flock is a sure sign of laziness or indifference on the part of the owner. It is caused by a parasite, and is, therefore, a contagious disease. When it first appears the shanks and toes become covered with a dry scaly substance which increases quite fast until it forms crusts so thick as to obscure entirely the original shape and color of the legs. It is most common among the feather legged varieties and spreads much faster in damp, filthy quarters than in clean, dry ones.

The treatment is very simple, but is also very effective. Apply thoroughly, with the fingers, some carbolized vaseline to every part of the shanks and toes. Repeat every two days until the legs are clean.

Lice.

THERE are several varieties of lice which infest the hen house. There is the common white or gray louse, which

sticks to the fowl day and night.   The same variety is found
on young chicks and is commonly called the head louse because found on the head and fastened to the skin like a leach.
Then there is the red louse, or red mite, which works only at
night.   During the day he will be found under or on the
roosting pole, or on the sides of the house.   He is bright red,
round and rather smaller than the head of a pin.   Frequently
the mites will congregate on a part of the wall so thick that
one would think the wall was covered with fresh blood.

There is also a brown louse, larger than the red and
not so large as the white.   The habits of this are similar to
both the others, that is to say, many will leave the fowl in
the day time and be found in the house, but some of the
more greedy will keep at work day and night.   This is the
kind that bothers the sitting hen the most.   Sometimes she
is compelled to leave her eggs, and, in such instances, one
looking into the nest will see no eggs there as they will be
completely covered with a mass of the dirty brown lice.

In this connection a very good answer appeared in the
notes and queries of a recent poultry paper.   The question
was like this: "What is the matter with my chickens, they
have such and such symptoms?"   Answer, "Look for lice,
and if you find them remove by doing thus and so.   If you
do not find any do just the same, for they are there, only you
do not know how to look for them."

The prevention and treatment are identical.   Keep
dropping-board clean in hot weather; sprinkle slaked lime
over it occasionally.   Have the roosts and dropping-board
arranged so that they can easily be removed.   Take them
out in the yard twice a month, in summer, and paint them
all over with kerosene, at the same time paint walls and
cracks near where roosts belong.   The same night go into
the house and sprinkle a little Lambert's Death to Lice over

the back of each hen. Clean out the nest boxes and paint inside and out with kerosene. Refill with clean nesting material and sprinkle a little Lambert's Death to Lice in it. Never set a hen without dusting both her and the nest thoroughly with the same powder, and repeat at least three times while she is sitting. When the chicks hatch, welcome them with a good dose of Lambert's, and repeat at least once a week, for the first two months of their lives.

If the chicks are badly infested with head lice, the quickest way to relieve them is to apply a very little vaseline to the top of their heads and under their wings. After one application of this the free use of Death to Lice will keep them away. Do not forget to keep the chicken coops clean, as filth is the very best place for breeding lice.

Gapes.

THIS is an affection seen only in young chicks from the third week to about the third month. It is, fortunately, not common in moderate climates, although said to be quite prevalent in the south.

Gapes is caused by the presence in the wind-pipe of one or more thread-like worms. These little worms attach themselyes to the lining membrane of the wind-pipe and cause it to swell so that it fills the whole caliber of the pipe and the chick dies from suffocation. The principal symptom is gaping. The chick stretches his neck and opens his mouth to its fullest extent. He does this repeatedly and soon refuses to eat, becomes dumpish, and, if not relieved, dies. The only preventive is absolute cleanliness about the coops and yards.

The treatment of gapes is not very satifactory. It consists in removing the worms from the wind-pipe. This can be accomplished by means of an instrument known as the gape worm extractor. The operation requires some skill and more patience. When a large number have to be treated the treatment is wholesale, so to speak, and the usual method is to smoke the worms out. The chicks are shut in a tight box, which is then filled with fumes burning of sulphur or carbolic acid. or with finely powdered slacked lime. The trouble with this method is that the worms will stand about as much as the chicks will, and you will be very lucky if you can stop at just the right moment, that is, when the worms are killed and before the chick are.

Better direct your energies to stopping the spread of gapes than to doctoring those already affected. Take all the sick and place them in a clean, dry coop, with sand and air-slacked lime on the floor. Take the rest of the brood and all the chicks that have access to the same yard, put them into quarters by themselves and watch very sharply, so as to remove each one to the hospital coop as soon as it shows a symptom. Be sure that any chicks that have not been exposed to danger are kept away from the infected yard, from the quarantined chicks, and, of course, from the sick ones, until the disease is thoroughly stamped out.

The infected coops and yards must be disinfected. A good way to do this is as follows: Burn all old coops that are not of much value; mix a hogshead of corrosive sublimate of strength of 1 to 2000; heat to boiling point enough of this solution to saturate every part of the coops. Sprinkle the rest of the solution over the ground. When the coops are dry give a good coat of whitewash. Sprinkle air-slacked lime over the ground until no earth can be seen. Leave alone for two weeks and then spade and sow down to grass. Put no chicks into this yard for two years. Fowl may be kept in it after the grass is grown, if necessary, but no chicks.

Pip.

THIS is a disease of young chickens and is practically a cold. It occurs oftenest in chicks whose parents have had roup, or have been inbred too much. It is sometimes caused by damp and filthy coops.

Treatment: Give dry, clean quarters, and wash mouth and nostrils with a week solution of chlorate of potash.

Chicken Pox.

THIS is a highly contagious disease which affects both old and young. It is rare now. It is characterized by black, hard warts or growths on the head and face.

The only treatment is to quarantine and keep the warts greased with carbolized vaseline. Fowls will generally recover and be as good as ever, while chicks almost always succumb within a week or two after they are taken.

Going Light.

THIS is not a very definite term, and the condition to which it is applied is also called consumption, scrofula, congestion of the liver and inflammation. It occurs occasionally in flocks that have the best of care, so it seems there is no sure way to prevent it.

It is undoubtedly a disease of digestive organs, possibly the liver. Autopsies often show a liver rather too large, but no other abnormal condition visible to the naked eye. The symptoms are great emaciation, extreme pallor of the face and comb, ruffling of feathers and general dumpishness. During the first of it the appetite is fairly good, but later disappears entirely.

When a disease attacks a chicken that is getting its second feathers, as it often does, it is, as a rule, fatal. To be of any avail treatment must be begun very early. Give sulphate of strychnine, 1-200 grain, three times a day, and color the drinking water with tincture of cloride of iron. Feed meat, green food and some cooked food, as bread or mash.

When the patient is a grown fowl the treatment is somewhat different. Shut up in the coop with clean sand on the floor, give calomel, 1-10 grain, every two hours for five times, and no food of any kind but plenty of water. The next morning, after these five doses, the droppings should be found in the sand, abundant and rather loose; if they are not, give a level teaspoonful of Epsom salts. After the bird has been well physicked in this way begin to feed soft food rather sparingly until your patient feels really hungry. Give the strychnine and iron, as in the previous case. As soon as the appetite returns put her back in the run where she can get more exercise and variety of food. Watch her carefully and if she grows worse again repeat the former treatment of calomel. It is often necessary to do this three or four times before thorough recovery takes place.

Now in conclusion, just a word. Remember that you will be well repaid for all the time and pains which you care to spend in giving your Bantams all proper care to keep them in good health. On the contrary, in nursing sick Bantams, your time will be frequently thrown away. The moral of this is: Do your best to prevent disease, and when it does appear, as it sometimes will in spite of your best endeavors, do not be afraid to use the hatchet.

INDEX.

BANTAM BREEDERS

From Whom You Can Purchase Good Stock and Receive Honest Treatment.

www.ingramcontent.com/pod-product-compliance
Lightning Source LLC
Chambersburg PA
CBHW060000230526
45472CB00008B/1875